Starting from Seed

Jack Kramer

Drawings by Robert Johnson

BALLANTINE BOOKS • NEW YORK

Library of Congress Catalog Card Number: 76-56132

ISBN 0-345-25502-X-150

Manufactured in the United States of America

First Edition: April 1977

Contents

Charts

1
A Seed Is Born

The houseplant on the windowsill, the vegetable in the garden, and the tree in the landscape seem to be a far cry from a seed. When we look at these full-grown specimens, we find it hard to think of growing large plants from tiny seeds. A seed is a miraculous gift of nature, an inexpensive gift that yields expensive and beautiful plants. Nature's gift is, also, fun. You can buy plants already grown, and not have to devote much time and care to them. But it is much more enjoyable to nurture a small seed into a plant—you can share the rewards of being a surrogate parent. Vegetables and fruits you grow from seed are far better and more nutritious than store-bought ones, and your own houseplants you raise are much more rewarding than those you buy.

There is not much mystery or trouble in sowing seeds. You need only a slight botanical knowledge of how a seed actually forms and grows. Once you have that knowledge, information about how to get seeds, and what types of plants to grow and how to care for them, you can start right in with this delightful form of gardening.

How Seeds Form

We think of flowers in visual terms because they are so beautiful to look at. But the main purpose of any flower is to produce seed.

A flower contains ten organs that eventually produce seed. All you really have to consider, however, are five organs: the stamens, the pistil, the anthers, the stigma, and the ovules. The stamens, the male organ of the flower, are made up of anthers, which develop and contain pollen. The pistil is usually in the center of a flower and is the female organ. The stigma is part of the pistil; the stigma receives the pollen grains from the anthers. The stigma is the resting place for the pollen grains as they germinate.

How do the pollen grains get from the anthers to the stigma? Nature solves this problem by giving flowers certain colors, nectars, or scents to which insects and birds are attracted. As a bird or insect lands on a flower's anthers, it picks up the pollen grains and transfers them to the stigma. If there are no insects or birds to move the pollen grains, wind and gravity will do it.

Once the pollen grains are on the stigma, they fertilize the ovules downward through a pollen tube. The ovules (inside the pistil) develop into embryos with surrounding coats; these embryos are the seeds.

To summarize, there are six basic sexual reproduction processes in the development of a seed:

1. Stamens and pistil (with their anthers and stigma) form within a flower.
2. The flower opens.

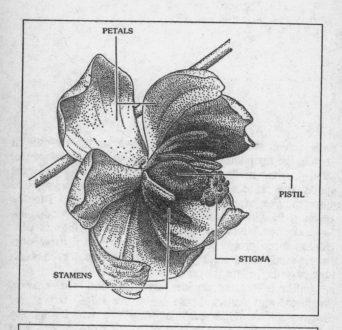

PETALS

PISTIL

STIGMA

STAMENS

THE REPRODUCTIVE SYSTEM

3. Pollen is transferred from the stamens (anthers) to the pistil (stigma); the pollen germinates; and the pollen tube is formed.
4. The ovule (egg) is fertilized.
5. The egg grows into an embryo with a protective coat: the seed.
6. The seed matures.

How Seeds Grow

Once a seed matures, to survive and germinate it needs the proper humidity, temperature, light, moisture, and growing medium. Nature endows seeds with several protective mechanisms to ensure their survival and growth. For example, a plant that drops its seed to the ground in winter has little chance then of reproducing, so these seeds have an ability to "sleep" during the winter until the proper growing season. Some of these dormant seeds need a long period of coolness, or alternating periods, before they will sprout.

Other protective devices are size, coatings, shapes, textures, and colors. Some seeds are large, but others are so minute (like orchids) that you can hardly see them. The size of a seed protects it from insects and animals. Seeds in cold climates have thick protective coatings; seeds in warm climates do not. Many seeds have beautiful shapes, and their surfaces may be grained or pebbled. Shapes and textures often disguise seeds in the garden, so predators will not notice them. Seed colors may be snow-white, jet black, or variegated. Color is another form of protective disguise for seeds.

Heat may cook some seeds, and cold with some frost may injure other seeds. Too much or too little water can make or break a seed (for example, vegetable seed needs tons of water). Hard, caked soil does not permit water to reach plant roots, and a soil spent of nutrients starves a plant. Without the right intensity of light, food plants cannot produce. In short, a seed survives and germinates only with the right conditions and care. But as you will see as you read this book, the care is not that difficult. A bit time-consuming, yes, but what parent has a completely easy job of rearing a child or, in your case, a seed?

Collecting Seeds

Gathering seeds from houseplants or roadside or garden plants does not harm the ecology, but tearing a complete plant from its natural place does. The trick to getting free seed is knowing when the seed pods are mature enough for picking. Immature seed will not germinate, and seed past the viable stage is also worthless. The seed must be ripe and then collected before it deteriorates. The best time for gathering seed varies from season to season, but usually it is when the plant's first seeds begin to fall, generally thirty days after flowers appear.

Some seeds are in pretty capsules or in pods. Strip these "containers" from the stem—there is no sense to cutting a branch. Then crush out the seed by hand, or tap the pods with a block of wood on a table. Taking seeds from plants in strangers' yards is not advised. However, almost all gardeners will let you collect a few seeds of their special plants *if* you ask first.

Once you have collected your seed from houseplants or roadsides or gardens, put them in a glass jar and

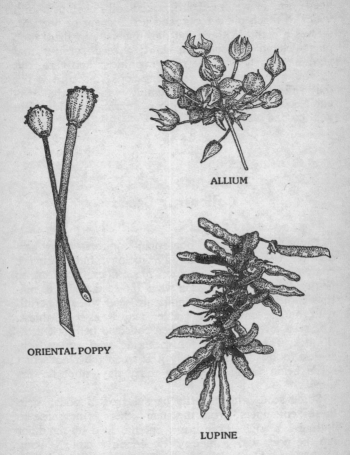

ALLIUM

ORIENTAL POPPY

LUPINE

SEED PODS

store the jar in a cool shady place if you are not going to plant the seeds immediately. Do not store seed too long (although most seeds are viable for several months), because you may forget them. Get seeds started right away!

The seeds of such fruits as grapefruits, avocados, and papayas can also be started. You will get lovely small houseplants from such seeds, but do not expect flowers. (See Chapter 6.)

Buying Seeds

There are more than five thousand varieties of flowers and vegetables for sale to home gardeners. Many of the flower and vegetable seeds are produced in California, where planting time is from November to May and harvesting is from June to December. The seed you buy is the result of years of cross-pollination, care, and study. Hybridists take infinite time to bring you the very best variety of a certain species. Cross-pollinating creates a variety of seeds with outstanding characteristics, such as color, form, flower size, or floriferousness.

Flower-seed plants are harvested and threshed in much the same way as grain—by machinery. (Some annuals and perennials are handpicked.) Plants are cut and then dried for ten to twenty days. After this the seeds are removed and are cleaned. Seed must be kept dry, so moisture-proof packets have been developed. All packets contain information about planting and starting the seed. You can buy seed in packages almost any time through the year, but the best time for sowing is in spring and fall, when the weather is good for germination. In the spring, warm weather is on the way; in fall, there is still time before very cold weather starts. This applies to seeds planted indoors too; if you start seed in March or April, by May or June warm weather permits transplanting seedlings outdoors.

Asexual Propagation

This book is about growing plants from seeds, a sexual propagation method. Basically, sexual propagation involves sowing, germinating, thinning, and transplanting seed. However, there are asexual methods of getting new plants. These methods involve taking parts of already-growing plants and growing them separately. Once you have grown some plants from seed, you may want to reproduce more plants asexually, to add to your garden and to save money.

There are five basic asexual propagation methods: offshoots, runners, cuttings, grafting, and layering.

Plants like chlorophytums, saxifragas, aechmeas, neoregelias, and orchids produce small babies, or offshoots, at their bases. These offshoots can be removed when they are a few inches tall and potted for new plants.

Using cuttings—stem, root, or leaf—is the easiest asexual propagation method. For example, a leaf cutting just involves snipping off the terminal growth of a mature branch, removing the bottom leaves, and leaving three or four leaves. Insert the cutting in a starter mix and give it humidity and water. When the cutting shows new growth, plant it in soil.

Grafting can be done by using the root, crown, or top of a plant. Tip, mound, trench, and air layering are other methods of asexual propagation. If you are interested in knowing more about asexual propagation (and you should be), consult any of the books listed in the bibliography.

2
Sow to Grow

Once you have collected or bought your seeds, you will want to get them growing. You can start your plants indoors or outdoors. Most people start seeds indoors to get a headstart on spring. Also, many plants need a long growing season to come into flower and fruit before frost, so starting seeds indoors gives them this season. Another good reason for indoor sowing is that, rather than planting only what is available in seed packets at local nurseries and stores, you can plant any vegetable or flower type you want. Finally, often indoor sowing is the only way of having a specific species or plant that may not be available commercially or that you are particularly fond of. Plants *can* be transferred outdoors when ready, but houseplants, vegetables, and some flowers can be permanent indoor residents.

To start your seed sowing indoors, you will need the right spot in the home, and containers and growing mediums. That is what this chapter is all about. The chapter will also cover starting seeds in small spaces outdoors (such as in a cold frame), direct ground planting in the yard or garden, and sowing seeds indoors under artificial light or in window greenhouses.

Where to Sow Seeds Indoors

Do not even think of sowing seed in your living- or dining-room window. You want to putter around and experiment with your seeds; think of the mess that would add to the main rooms of your home! Instead, start seeds in the basement, kitchen, bathroom, or pantry—any area that has room and is more or less out of sight of visitors. The growing area must have ample space for containers and watering cans and not interfere with everyday living. On a less esthetic note, growing seeds in one central location makes tending and watering seeds easier than running around from room to room.

You do not need too much space for your plant nursery, just 5 or 6 feet in width and 2 or 3 feet in length. Select a place that is waist high so it is easy to tend plants, and where conditions are right for seedlings (although this can be modified somewhat): the growing spot should have a constant temperature rather than a fluctuating one, and some natural light but not direct sunlight. (If you want to start seeds under artificial light, you can use an unused closet, a dark nook or cranny, or any area with some open space and easy access.) Finally, if you can, have the growing area near a water tap.

Containers

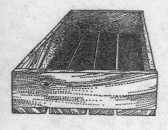

WOODEN FLAT

EGG CARTON

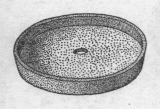

CLAY DISH

PLASTIC TRAY

CONTAINERS

Your seed containers should be small enough so they are easy to move and large enough to accommodate a minimum of ten to twenty seedlings. The wooden boxes (flats) nurseries use are good, if you can get some. Chances are you will not be able to find any flats, so use plastic trays. These containers, about 4 inches deep, 16

VARIOUS STAGES OF PEAT PELLETS

SEED CUBES

inches wide, and 20 inches long, can be purchased at suppliers.

You can also use such household containers as aluminum frozen-roll cartons, milk cartons sliced lengthwise, one-pound coffee cans, bread pans, and plastic egg cartons. (All these containers will supply space for few seeds, of course.) Be sure any household container you use has drainage holes at the bottom, so excess water can escape. You can also use standard plastic or terra-cotta pots; the shallow pots are the best ones for seed sowing. You can start four or six seeds in an 8-inch pot.

A convenient way to start *vegetable* seed is to use the special seed devices like Jiffy Pellets or ready-to-use peat cubes that contain fertilizer (the cubes are sold under various trade names; see Chapter 7, pp. 60–61). The pellets are compressed peat disks; you just add water. Fiber seed trays and pots are also available. (We discuss these starters more fully in Chapter 7.)

Growing Mediums

Any growing medium for seeds has to be sterile or the fungus called "damping off" may strike. (Damping off is discussed in the next chapter.) Nurseries carry the five most popular sterile starting mediums: milled sphagnum; vermiculite; perlite; peat moss and sand; and a mixture of equal parts of vermiculite, sphagnum, and perlite.

Milled sphagnum is one of the best starting mixes; just remember to keep it evenly moist. Vermiculite is expanded mica that retains moisture quite a while. Packaged vermiculite has added ingredients in it. Perlite is volcanic ash. By itself perlite tends to float and disturb the seed bed, but if you mix it with a bit of sterilized soil (all packaged soil is sterilized), you will have a good starter. Perlite will also retain moisture, which is what seeds need.

Try to avoid using packaged soil mixes, because they are too heavy for seeds. If you have to use them, combine them with some fine and porous sterile soil.

Moisten all growing mediums a few hours before you use them, or you will have a mess on your hands. Finally, keep your growing medium always moist (but not soggy) while seeds are growing—uniform moisture is the most important factor of successful seed sowing.

Starting and Growing Seeds

Fill the containers with a premoistened starter medium to within 1/4 inch of the top. If the planting bed is lower than 1/4 inch, the growing medium is not exposed to enough good ventilation; proper ventilation deters damping off.

If you bought seeds, open the packets at one end by making a diagonal cut. Tap out the seeds into the palm of your hand and plant them. Dustlike seed will be inside a smaller envelope; slit this inner envelope and let the seed fall out one by one over the growing medium.

Both large and small seeds need to have space to grow, or you will crowd them and thus encourage damping off. Do not brood about how far apart you plant seed; just scatter the seeds with some space between them.

Once the seeds are scattered on the growing medium, they have to be either imbedded into the medium or

SOWING SEED IN FLATS

1. Fill with sterile growing medium and moisten.

2. Sow seeds according to direction on packet. Cover seeds lightly with soil and mist.

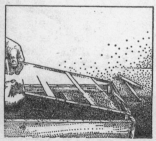

3. Cut dowels to lengths of one-half inch from top of flat; insert dowels in corners on both sides of flat. Top with glass panel or plastic covering to ensure humidity.

4. Remove seedlings for transplanting.

POT 'N' POT METHOD

1. Fill large pot one half full with growing medium.

2. Place smaller pot in the center and add more soil.

3. Sow seeds according to directions on packet.

4. By filling the center pot with water this method allows for uniform moisture to reach the seeds.

lightly pressed in. This depends on the type of seed. Fine seed such as petunia, begonia, and snapdragon should be lightly pressed into the starter. But large seed like morning glory should be imbedded into and covered with the medium so they are not seen. Furthermore, tree or shrub seed with very hard coating should be nicked first before planting. (Chapter 9 will discuss how to nick seed.)

Now that the seeds are in place, water them. Do not just dump a glass of water into the container, because the seeds will float all over. And do not use a mister bottle to moisten the medium; you can literally blow the seeds out of the medium with a jet-propelled water spray. Instead, water from the bottom, or use a watering can with a "rose" on the end. To water from the bottom, soak the containers in a sink filled with water; slowly submerge the containers into the water, then lift the containers out of the water.

We said in a previous section that the most important factor of seed germination is continual moisture. A good way to keep the medium moist without overwatering (which most people tend to do) is to cover the tray with some clear plastic set on four sticks, a stick at each corner of the container. Be sure the plastic does not touch the medium. Punch a few small holes in the plastic to ensure that ventilation gets inside; otherwise too *much* moisture will accumulate inside the "tent." Once leaves develop, remove the plastic.

While your seeds are growing, they should be in a bright but not sunny or dark spot. Maintain a constant temperature of 75 F.; most seeds germinate at this temperature. If it is difficult to maintain a constant temperature, use heating cables. (These are electric cables purchased at nurseries.) Put the cables at the bottom of the container before you put in the growing medium; or at nurseries buy wooden containers with cables attached. (This wooden container with cables is really quite handy; lately I have been using one for sowing seeds.)

As they are germinating and growing, most seeds will not need feeding. However, some annual and perennial and tree and shrub seeds grow slowly after germination and must remain in their containers for many weeks. These seeds will thus need a very weak solution of plant food. Use Rapid Grow or Hyponex, but one-half strength rather than full strength as prescribed on the packages. Use the food about once every ten days until seedlings are ready for transplanting.

Seedlings are ready for transplanting when they have at least four leaves. Once the seedlings are transplanted, you will have to follow a new set of rules. But do not worry about that just yet; we will consider transplanting in the next chapter.

Seeds Under Lights

There is a proven reason for the current craze of growing plants under artificial light. Fluorescent or incandescent light helps produce healthy plants and somewhat speeds up the seed-germination process. You can use any type of fluorescent lamp, including daylight and warm white, rather than a special grow-type lamp.

You can try just two 40-watt fluorescent lamps to a setup, but I have found that seeds do better under a combination of lamps. I use the two fluorescent lamps with four 8-watt incandescent lamps. Recently a combination fluorescent and incandescent lamp in one tube has been introduced; you might want to try this lamp.

Put containers 6 to 8 inches below the light source. Leave lamps on twelve to sixteen hours a day. Plants under artificial lights grow continuously, so you will have to feed the plants with a weak solution of 10-10-5 plant food as leaves appear and start to grow. As with sowing seed in regular light, maintain a constant temperature, good humidity, and a constantly moist medium that is not too soggy or too dry.

Window Greenhouses

Window greenhouses, for inside growing, attach to any type of window, large or small. You can get glass-and-wood or glass-and-metal greenhouses, or the one-piece molded injection acrylic ones. You can also make your own window greenhouse. In most cases you will not have to remove the existing window sash; any window greenhouse fits into the existing window as it is. The beauty of the window greenhouse is that the existing window can be opened or closed at will, so you can control the ventilation. Inside the window greenhouse, humidity and temperature stay constant, and this is the ideal way of starting seeds.

If your window greenhouse gets somewhat cold at night, then use heating cables. You can also install artificial light at the top of the window greenhouse, if you think it is necessary. However, in window greenhouses seedlings usually get too much natural light, so you may have to apply a covering to block out very strong sunlight.

GROWING SEEDS OUTDOORS

1. Preparing soil and rows.

2. Sow seeds in mounds and label.

3. Remove weeds as seeds grow.

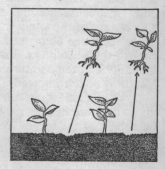

4. Thinning.

Growing Seeds Outdoors

Starting seed in the garden requires different methods than when you start seed indoors in containers. Many vegetables, annuals, and perennials do better when planted in open ground than in indoor sowing. Nasturtiums, cosmos, and zinnias are examples of such flowers; beets, carrots, and lettuce, which germinate well in cold moist soil, represent the vegetable group. In addition, these plants and some varieties of peas just don't transplant very well, so this is another reason for direct planting.

The soil for direct seed sowing should be carefully prepared as directed in Chapter 3 (see "Preparing the Outdoor Soil Bed").

Once the soil is prepared, you can sow seed—and this is usually done in rows. Space vegetable seed according to the table in Chapter 7. It is little problem to space large seeds such as radish and spinach for example, but small seeds can be tedious to sow.

The row method of planting gives you space to walk between the plants, so you can tend them easily; and row planting also helps you differentiate between weeds and seedlings when the seeds germinate. (And there will always be weeds.) The distance between rows of plants varies and is in the table in Chapter 7.

Before putting seed into the ground be sure the soil is moist. The shallow trenches between the rows tell you when you have watered enough; the trenches will show signs of water. Label each row so you will know what you are growing. Even the experienced gardener is apt to forget what is where.

Keep the seed bed evenly moist but never soggy during the crucial time of germination. And do remove weeds when they appear; otherwise, they sap strength from the soil that your plants need.

When the seedlings are about 2 to 4 inches tall, they must be thinned so there will be ample space for them to grow. This means pulling up smaller seedlings and discarding them. Or you can re-use the seedlings by giving them to friends or for fill-ins in other areas.

When you thin plants—no matter how carefully—you will disturb those left in the ground and they will suffer somewhat. Furnish plenty of water before and after thinning. Not all seeds will come up; some may be sterile, others just don't germinate, because of lack of proper conditions. But no matter—you'll have enough seedlings from a standard package of seeds to grow them on.

Cold Frames

If you have very little indoor space and you have some (3 × 3 feet) small outdoor area, consider a cold frame. A cold frame is essentially a very low greenhouse; it is a glassed-in box, set into the ground. The cold frame gives you a chance to start seeds that need alternating periods of freezing or thawing to break the dormancy that seeds start into in the fall or winter. Or you can use the cold frame to start seeds in early spring, before the weather is settled.

You can buy commercial cold frames at nurseries. Or try the one-piece molded acrylic bubbles now available. But even easier is to make your own cold frame from four boards and an old glass-paned window sash.

DISCARDED WINDOW SASH   COMMERCIAL COLD-FRAME

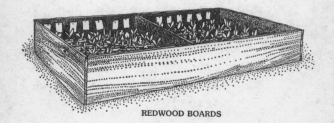

REDWOOD BOARDS

COLD-FRAME SOWING

Merely nail the boards into posts set into the ground, and put the sash on top. Pitch the sash slightly, so water rolls off. Bury the base of the boards about 6 inches into the ground; using 2 × 12 boards for the frame allows sufficient growing space. (Some commercial cold frames have bottoms; some do not.)

The cold frame will need some shading, so build a lattice over the frame, or put it in a spot that is out of the sun. An excellent addition to a cold frame is to hinge the top—the top can be propped open as necessary to ensure good ventilation for the seeds.

(Start your seed in the cold frame in the soil in the bed. Or start it in containers, as described for indoor starting, making sure you keep the growing medium uniformly moist and use a *sterile* medium. Sometimes containers, themselves, are set in cold frames.)

3

Growing On

Through the years, I have found that most gardeners lose half their plants at the crucial stage of transplanting. After seeds germinate and start leaf growth—indoors—a new set of cultural rules is necessary. You have to transplant seedlings from their original container into smaller but individual ones before they can be planted outside in the ground. You have to use the right soil and see that the plants get the right amount of light. You have to acclimatize the plants and slightly prune them. These and other rules have to be adhered to; they are vital parts of good plant growing.

Transplanting

You now have a good bunch of seedlings up and ready to grow. In fact, you now have too many seedlings, all vying with each other for growing space and air. At this point it is time for transplanting.

27

TRANSPLANTING SEEDLINGS TO POTS

1. When seedlings are about two inches high and have separate leaves, remove them for transplanting.

2. Take as much of the root ball as possible and place seedlings in small pots.

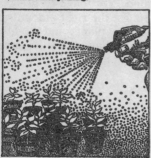

.3. Keep the seedlings moist with a fine-mist spray.

4. After a few weeks, transplant to larger pot and place in permanent position at a window, or outdoors if for the garden.

Transplanting should be done only when the first set of *true* leaves—four leaves—show. To remove the seedlings, use a blunt instrument like a wooden ladle or the handle of a small teaspoon. The idea is to gently take out the seedling with as much of its root ball intact as possible. If you brutally yank out the seedling, it will suffer great shock and probably die. So gently lift out the little plant.

First, however, prepare the smaller, individual new containers with a mixture of soil, sand, and compost in equal parts. (Compost is decayed organic matter, like decayed leaves, that you can buy at nurseries.) Fill the containers to 1/4 inch of the top, and lightly water so the soil mixture is moist. Using the same ladle or spoon handle you will use to lift out the seedlings, make a slight indentation in the soil mixture in each container to accommodate the seedling roots. Set the plants in these pockets up to their collars; tuck the soil in place firmly but not too hard. Water the soil immediately after transplanting, and put the containers in a bright but not sunny place for about a week to give the young plants a chance to recover from their move. Then move the plants into bright light.

Care of Transplanted Seedlings

Now that the little plants are in new containers, you must keep them growing before moving them outside or putting them in permanent indoor places. It is absolutely vital that the growing medium is evenly moist at all times. To see how moist the soil is, feel the top of it with your finger. If it is moist to the touch, fine. If it is not, water the soil immediately. When watering, do not dump water into the container. Use a light mist of water; apply the water gently and evenly over the surface. Use tepid water, never icy-cold or hot water.

To help encourage growth, lightly feed the fledglings with a mild plant food. Apply the plant food about once every five days. Also during this all-important adolescent stage, watch the new plants. If they are getting leggy, pinch them back to encourage branching and busy and compact growth. To pinch a plant, nip off the top growth of the stem with your thumb and forefinger. This pinching is especially helpful to single-stemmed plants.

Hardening Off

You have transplanted your seedlings and are caring for them. Now they need to be acclimatized if they are to be grown in the garden. Remember, your plants have never been exposed to extreme outside conditions like full sun or wind, so they have to be rehearsed for their outdoor performance. This period of acclimatizing, is called hardening off. You have to put the seedlings into a protected area—porch, back steps—for a few days and nights before planting them in the garden. Be sure the plants are out of direct sun but getting some bright light. In a few days, move the plants into stronger light and a more exposed area and keep them there for a week or so. The plants are now ready for transplanting into the garden.

While the plants are hardening off, be sure to keep the growing medium uniformly moist.

If some nights are *very* cold, move the plants back into the house, or they may die. When the weather is settled (late April in most areas), and the plants have hardened off, get ready to transplant them into their permanent outdoor spots. (Seeds to be grown on indoors can be placed in light immediately.)

First prepare your soil bed, as outlined in the next section. Cut and remove each plant from its container with an old bread knife or mason's trowel. (Do this gently.) Dig holes deep enough to hold the plant's root ball and wide enough to accept the plant's circumference. Place all the plants into position; mound the soil around their collars, and smooth and level it down with your hands. Tamp down the soil firmly but not hard; you want to eliminate air pockets in the soil, but you do not want to compact the soil so that air and water cannot enter.

Water the plants thoroughly, and keep them well watered. Unless it rains, they will need watering about every third day.

Preparing the Outdoor Soil Bed

Some people just turn over the outdoor garden soil and plant their seedlings. But it is much wiser to prepare the soil bed so that the new plants will have adequate nutrients. (It is senseless to grow seedlings indoors and then transplant them to an unfit outdoor soil.) This preparation is called soil conditioning.

A few weeks before you plant the seedlings outside, start preparing the soil bed. Dig down 15 to 18 inches, and turn over and pulverize the soil so it is porous and mealy. Do this job with a spading fork or a shovel, or rent (by the hour) a small garden tiller. Most likely your outdoor soil will be drained of nutrients, so enrich it with leaf mold, or compost, and some dried cow manure. Again rake over the bed so it is porous and mealy, not too fine or filled with clods of soil.

If you are growing vegetables, remove the top 12 inches of soil and replace it with fresh topsoil; then add the manure and compost.

It is a good idea to test the soil's pH; that is, its acidity or alkalinity. Too high or low a pH can lock certain elements into the soil so that plants cannot get them. A soil is neutral—neither acid nor alkaline—if its pH is 7.0. Most plants like a neutral (7.0) soil or one slightly below (6.5, or more acid). You can test the pH of your soil with one of the inexpensive soil-test kits advertised in garden magazine ads. The results will give you an idea of what your soil needs in terms of phosphorus, lime, nitrogen, and so forth. When the time comes, you will know what plant foods plants need, rather than just guessing.

Indoor Seedlings

So far in this chapter I have talked about starting plants indoors and transplanting them outdoors; but many people, especially apartment dwellers, do grow herbs, vegetables, and houseplants indoors—and keep them indoors. I myself have grown tomatoes, cucumbers, and green peppers as well as cut flowers like petunias, marigolds, and dahlias inside. Just what you can grow indoors depends on your individual conditions. If you have a sunny window, try a few vegetables and herbs; if you have a window greenhouse, grow some annuals and perennials. Houseplants can be grown indoors all year, of course.

Indoors plants do not have to be hardened off; they can immediately go into bright locations that have average home temperatures. If the proper amount of humidity is lacking, cover the new plants with a Baggie for a few weeks to be sure they get off to a good start. Use clay pots or small window boxes as containers.

Keep the seedlings evenly moist, and always use a fertile and porous soil for them. Packaged soils are fine, but do add some compost and manure to them. I use a 4-cubic-feet bag of standard houseplant soil and add two cups of compost and one cup of manure to it. But feed your indoor plants judiciously. Some plant food is necessary in the first few months, and when the plants are really growing, you can increase the feeding somewhat. I use a 10-10-5 plant food for all plants.

Damping Off

Fungi cause damping off; the fungi may be harbored in the soil, or they can be carried to the plants via water. A plant has damping off if the stem rots and/or has a slightly gray mold on it. Sometimes seeds themselves get damping off; they rot before they can even start growth. Damping off is the universal nemesis of seeds. (Similar to damping off, but caused by cultural conditions, is seeds' not germinating, or dying after they sprout. This can be caused by drying, high temperatures, or cloudy days coupled with moisture at night.)

The best method of preventing damping off is, as previously mentioned, to use sterile starting mediums like vermiculite and perlite, because fungi have little chance of starting growth in such mediums. Good sanitation—keeping the growing area spotless—is another way of thwarting the disease. Any decayed or diseased plants or organic matter should be discarded immediately. Do not keep "rotten" plants in the growing area.

If you keep the growing area clean, make sure the growing medium is never excessively moist, provide good ventilation, and use containers with drainage holes, the chances are your seedlings (or seeds) will not be sick from damping off. But if the disease does hit your plants, you will have to use a fungicide to eliminate the problem. There are many fungicides available at nurseries. Follow the directions on the package or bottle to the letter, and keep all fungicides out of the reach of children and pets.

4

Caring for Mature Outdoor Plants

Once your seedlings are planted and growing outdoors, you have to care for them with the correct watering, feeding, and harvesting, and be sure they receive correct lighting. The care of mature plants is somewhat different from the care of young seedlings discussed in the previous chapters. Mature plants need minimum care to prosper, but still they need *some* overseeing.

Light

Plants generally need good light—sun—to mature into beautiful specimens. Plant most annuals and perennials and vegetables where they get at least four hours of sun a day. If your property is heavily shaded, grow wildflowers or shade-loving annuals such as impatiens, and lobelia. And perennials such as candytuft and forget-me-not.

If your annual and perennial garden is for cutting, you of course want maximum flower production. In this case a south or east exposure is a necessity. These exposures provide the most sun; the most sun in turn provides the most flower production.

Most vegetables also thrive in south or east exposures, but vegetables like spinach and cabbages will succeed in a west garden as long as they get a few hours of sun daily.

Watering

No matter what kind of plants you are growing, most of them need lots of water. If the region you live in has adequate rainfall, your watering chores will be minimal, but in many areas you will have to water plants at least a few times a week. Just how many times depends on the rainfall in your region. Generally, water in the morning because then soil has a chance to absorb the moisture. Dusk is the next best watering time. Nighttime is not good for watering plants, because when temperatures are cooler the plants do not absorb the water.

You must water thoroughly; the water must penetrate the soil and get to the roots of the plants. Surface watering can actually harm plants, since their roots will reach upward to get moisture. Thorough watering ensures a natural growth of roots and thus good plant growth. The soil should be evenly moist throughout its depth, never dry or soggy. If the soil has good tilth (porosity), your plants can take buckets of water.

With a hose it is almost impossible to give plants

enough water; only a sprinkler can do this, since it takes about two hours to penetrate 20 inches of soil (assuming the soil has good tilth). A slow, steady shower of water from a sprinkler is the most efficient way of watering your plants. There are several types of sprinklers available. I cannot recommend any one of these. Just buy the best sprinkler you can, and do not bang it around or drop it, because any sprinkler is a precision instrument.

Feeding

You should moderately feed annuals and perennials, vegetables and herbs. But feed them only when they are mature enough to absorb the additional nutrients—usually when the plants have been in the ground for two or three weeks. Start a feeding program with a basic 10–10–5 plant food (not too strong or too weak), about every other watering, through the growing season of early spring into mid-summer. Feeding plants in late summer and early fall accomplishes little since in most areas the growing season is over by Labor Day. In all-year temperate climates, feeding can be extended into the early fall.

As mentioned, I prefer a basic 10–10–5 plant food. Plant foods vary greatly and come in many strengths; so as a start, use the 10–10–5 food. Occasionally I use a specific plant food, such as vegetable plant food for vegetables. You too might eventually want foods designed for specific plants. Just what you use depends a great deal on what you grow. Generally, it is better to use a mild plant food as described than, say, a plant food such as 30–20–10, which can harm some plants. Potent plant foods can cause leaf burn on some plants.

Harvesting

After you have started vegetable plants and nurtured them to bearing stage, you want to pick the produce when it is its tastiest and best. To harvest your decorative plants—annuals and perennials—be sure the flowers have opened to the fullest, and cut them early in the morning. Take as much of the stem as possible, and immediately plunge the cuttings into water.

Chapter 7 gives schedules for harvesting produce from your vegetable plants, an important part of gardening.

5
Houseplants

Growing your own houseplants from seed not only saves you acres of money, but it enables you to grow a specific species (too often today suppliers carry only what is popular). Small cacti and lovely gesneriads, ficus trees and begonias, ferns and palms and other plants provide you with fine indoor accents. There is a vast array of houseplants you can start from seed and grow on to maturity—and what satisfaction in saying "I grew it myself."

The procedure for seed-starting houseplants is the same as for most other plants. Let us now discuss the various houseplants you might want to start. The following list is by no means complete; it is a selection of houseplants I myself have grown successfully from seed. Seed of the following plants can be purchased in packets from suppliers.

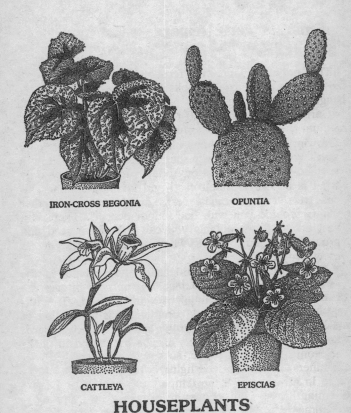

IRON-CROSS BEGONIA

OPUNTIA

CATTLEYA

EPISCIAS

HOUSEPLANTS

Emerald Fern (Asparagus sprengeri)

This popular plant with fernlike foliage (it is from the lily family, and only looks like a fern) is ideal for hanging baskets.

Germinate the seed at 70 to 75 F.; keep the growing medium evenly moist. Germination starts in twenty to thirty days. Transplant the seedlings when they are 4 to 5 inches high.

Begonias

There are many types of begonias, including the rhizomatous, hairy-leaved, angel-wing, and rex types—which four are hard plants to find at nurseries, so it makes good sense to start your own from seed if you want them.

Begonia seed is very fine, so carefully sow it in seed containers. Cover the seed lightly with the medium, and place in a spot where the temperature is at least 72 F. Germination usually occurs in fifteen days, although sometimes seeds will germinate in just one week. When the seedlings are a few inches high, transfer them to larger containers, and in another few weeks transplant them in permanent pots of rich soil containing some peat moss. Watch for mildew.

Cacti

This very large group of easy-to-grow plants includes unparalleled gems: parodias, lobivias, mammillarias, and dozens of others. Some cacti are available at nurseries, but the most interesting species are not, so start the plants from seed.

Cacti seed is fine or large. Do not cover any type of cacti seed; just sow the seed on top of the vermiculite or starting mix. The seed needs temperatures of 80 F. Some cacti germinate within fifteen days, but others take about thirty. Keep the starting mix just moist, never soggy. When the seedlings are up, transfer them to a larger container of equal parts soil, sand, and gravel, and grow them on a few weeks. Then transplant the plants into individual pots of the same mix. Apply a fertilizer in about one month.

Ficus

The Ficus family includes the favorite *Ficus benjamina* and rubber tree. Plants are generally easy to start from seed, but unlike most seeds, ficus need coolness—about 68 F. Imbed the seed into the mix, and keep the mix uniformly moist. Germination starts in twenty days. Ficus plants grow quickly, so transplant them every two months into rich soil until they are about six to eight inches tall. Then put the plants in permanent pots.

Gesneriads

This large and lovely group of plants includes a host of favorite indoor subjects: African violets, kohlerias, columneas, smithianthas, streptocarpus, and others. Most gesneriads are characterized by somewhat furry leaves, and all plants have beautiful flowers.

You can start the seed of most gesneriads in 75 to 80 F. Keep the growing mix uniformly moist. Press the seeds into the growing soil lightly, so they are exposed to the light. Do double transplanting; that is, transfer the seedlings when they are a few inches high, and then transplant them when they are about 6 to 8 inches high into permanent pots. Germination time varies. For example:

Columneas	15 to 20	days
Episcias	25 to 35	days
Gloxinias	15 to 20	days
Rechsteinerias	20 to 30	days
Saintpaulias	20 to 25	days
(African violets)		

Geraniums

These are the colorful plants that people either have great success or great failure with. Unfortunately, I fall into the latter category, but perhaps you will benefit from my mistakes.

Germination time for geraniums is erratic. I have had some seed sprout within twenty days, and others not show growth for months. Unlike most houseplant seeds, geraniums like coolness, so start the seed at 65 F. When the seedlings are up, put them in rich soil in 2- or 3-inch pots; in a few weeks, transplant them into individual containers to grow on. Do not give up; geraniums are worth the extra time and effort because of their fine flowers.

Orchids

Of all the exotic plants, orchids are perhaps the number one houseplant in many regions of the country. However, orchids are generally difficult to grow from seed and take expertise and patience. But if you have an orchid plant and want to try your luck—and it is very satisfying to get orchid seeds growing—do so. To pollinate orchids, use an artist's paintbrush and transfer the pollen from the stamens to the pistil. Leave the pods on the plant until they naturally crack open. (See Chapter 11 for a discussion of hybridization.)

Orchid seed is fine, almost dustlike, so just sprinkle it on the surface of a suitable growing medium such as peat moss or osmunda (available at nurseries). Maintain a temperature of 70 to 75 F. Water the seeds from the bottom because top watering washes them out. Treat the seedlings gently, and when they are ready for transplanting, plant several to a pot in osmunda or finely crushed fir bark. Orchids need excellent ventilation and almost perfect drainage.

Palms

These graceful plants are expensive if you buy them. You must wait a few years for a 3-foot plant if you grow seeds, but the effort is worthwhile. Sow the seed in shallow trays or pans of vermiculite or sphagnum at 65 to 70 F. Seed germinates irregularly: in sixty days or several months. Cover seed with 1 inch of medium. The palm family includes kentia palms (the most popular), rhapis, phoenix, and chamaedorea.

Ferns

Starting ferns from seed (actually spores) is a fascinating and pleasurable way of growing your own. Even under most adverse conditions ferns will produce seed spores indoors. Spores are the minute oval bodies on the undersides of fern leaves; spores look somewhat like scale insects. In some ferns the spores are on the edges of the leaves; in other ferns the spores are spread over the entire surface of the back of the frond.

Gathering Spores

Fern spores are ready for harvest when they are brown and slightly dry. Turn over the frond and examine them; the gathering time varies, so you must learn by close observation. If you take the spores before they are ripe, they will shrivel up and die. If the spores are left too long on the leaves, they will also shrivel and die. So it is wise to take spores at intervals, to be sure of getting some ripe ones.

Collect the spores by rubbing them from their casings. Do this with your fingers. Keep a sheet of white paper underneath as a gathering "pan." Here are the two different ways I sow fern spores:

Sowing Method 1

At the bottom of a shallow clay pot, put in broken pieces of pots (shards) and some pieces of brick rubble the size of quarters. Brick soaks up more moisture than ordinary crockery, and the spores can draw on this absorbed water to tide them over if you forget to keep the medium moist. Now put equal parts of sphagnum, humus, and sand into the pot. The starting bed should be 1 or 2 inches deep. Sow the spores on the surface of the medium. Cover the pot with a Baggie or a pane of glass. Do not water the seed from overhead; stand the pot in a saucer or tray to water from the bottom. Always empty the saucer by evening to prevent mildew and overwatering. But you must keep the medium uniformly moist, because if spores dry out they will die. Keep the temperature at a constant 75 F.

Some spores will take a few weeks to sprout; others will take several months. Keep the spore pot in a shady (but not dark) place. When you see the first flush of green on the surface of the medium, you are looking at the beginning ferns, or prothalli. This green fuzz is a one-celled body with rootlike hairs that fasten into the soil. One cell after another develops until there is a heart-shaped form about 1/4 inch in size. It is from this form that the new fern sprouts.

As the seedlings grow and start to crowd each other, thin them to about 2 inches apart (discarding some) so they have space to grow. In another few weeks remove each seedling and pot it separately in a rich porous soil with lots of sphagnum. Place the pots in bright light, and keep the soil evenly moist.

Sowing Method 2

You can also start ferns on bricks. This is an old method, but it works fairly well, sometimes better than a growing medium. Place common house bricks in a pan of water to within 1/2 inch of the top of the brick. Keep the pan filled with water. Lay sphagnum moss on top of the bricks, matting it in place as securely as possible. Put the fern spores on the moss. Grow the spores as in Method 1. Keep the container covered to ensure good humidity.

Later, you can pick the individual fern seedlings out with a stick, for thinning and transplanting.

6

Fruit Plants

We are all familiar with the seeds from grapefruit or oranges and the larger seeds (usually called pits) from avocados. All of these can provide pretty houseplants for the home—but don't expect fruit, for cross-pollination is necessary with most plants. Apricot and peach pits too will grow indoors, but these really need outdoor conditions for best results.

In this chapter we are mainly concerned with seeds of other common edible plants such as citrus and avocado and pineapple. Although pineapple is started from a top rather than a seed, because it is so popular I thought I would include it.

Avocados

Avocados are sprouting in homes faster than nasturtiums in gardens these days and rightly so. Once you eat the delicious avocado there is no reason not to plant the pit, because it yields (with little effort on your part) a leafy plant for the home. There are as many ways to start avocados as there are recipes for apple pie. I have even had an avocado pit sprout on a window sill without soil or a pot. However, for best results start your avocado this way:

Clean the pit in some tepid water, cut a thin section from the apex and the base, and peel away the papery pit covering. Now put the pit base—the broad portion—downward in water or in soil. If in soil, do not imbed the pit too deeply; cover it with about 1/2 inch of soil. If in water, find a jar with a mouth that will hold the pit snugly so that one half is in water, the other half out of it. You can use a large jar and use the toothpick method. Insert four toothpicks around the center of the pit to hold it so that one half is in water. When growth starts—and this usually happens in a week or two—and when there are four or six leaves, replant the pit into soil in a pot that is 2 inches wider than the pit.

Be sure the avocado has good drainage; although it is not choosy about soil—any houseplant soil will do—it is particular about stagnant water at the roots, which may rot the pit. Because avocados have the tendency to grow straight upward, it is wise to clip off the top once it is 12 to 16 inches high in order to encourage branching. This may or may not be successful, so to assure a leafy pretty plant start two or three pits to a large pot. Stake the plant; that is, supply a thin wooden stick into the soil next to the stem and tie the stem to it so the plant does not become ungainly.

Repot the avocado every six months, each time putting it in a slightly larger pot. Eventually you will have an attractive tree that will outgrow the windowsill. In that case, move it into the living room or kitchen, where it can be a fine decorative plant—for free.

POMEGRANATE   GRAPEFRUIT

AVOCADO   PINEAPPLE

FRUIT PLANTS

Citrus and Stone Fruits

You can, of course, buy attractive leafy citrus plants at nurseries, but they are costly. You can also start your own from the seeds of grapefruit, oranges, or lemons.

Start citrus seeds directly in soil in standard pots. Allow two seeds to an 8-inch container. Use a packaged houseplant soil and bury the seed half its depth into the soil. Water well and place the citrus in good light at a window. Seeds generally germinate in a few weeks, but don't give up—sometimes it takes a bit longer.

Once the plants are growing, apply a 10–10–5 plant food every other watering. When the seedlings are touching each other it is time to remove each one into separate quarters, so you can have at least two plants growing. Keep the root ball intact as much as possible when transplanting them, and prepare pots of soil for the plants. Set them in place and firm the fresh soil around the plants. Water them thoroughly.

When your plants are 6 to 8 inches tall and growing well, mist the leaves with tepid water every few weeks and also wipe the leaves with a damp cloth to keep the pores open. Remember, plants breathe through the leaves. Also, red spider has a fondness for citrus but frequent misting will keep it away.

It may take several years to get a sizeable plant from a citrus seed—three or four years—but while it is growing the citrus is always attractive, leafy green, lush, and branching. And since it costs little, growing citrus from seed makes good sense.

As mentioned, the stone fruits such as peach or apricots really need outdoor conditions to get them growing.

Indoor planting is rarely successful. Outdoors the stones or pits should be buried in the ground until February. Then uncover them and scarify (make a notch in the outer shell); then replant them 2 inches deep in a bed of sandy soil. In the fall, cover the seeds with about 6 inches of leaves. When the seedlings are about 12 inches high, they can be potted individually in containers or replanted in the ground. This is all a great deal of work, so here is one time I opt to buy prestarted small stone fruit trees if you have a hankering for them.

Pineapples

The pineapple is neither a pit nor a seed but I include it because it is so easy to grow and is so popular, ranking second to the avocado. Starting a pineapple is simple. Slash off the crown of the fruit (the part with the leaves) about one inch below the base of the leaves. Let it dry overnight to prevent rotting of the stem after planting. Now place the crown in sandy soil in an 8-inch container. Set the pineapple-to-be in a warm (75 F.) place where there is bright light. Provide good drainage and keep the soil just barely moist. Heavy watering is not necessary.

As new leaves appear on top, clip off the old dead ones from the original pineapple. Within a few months the plant will be attractive and handsome.

Pineapples are easily started, and the only problem is that occasionally the crown rots; to avoid this dry it as previously mentioned and be sure it is not kept excessively wet.

Pineapples are rarely bothered by insects or drafts or any of the more common plant diseases. Try one—it does make a handsome plant for indoors.

Exotic Fruits

The papaya and pomegranate are following the route of the avocado and many people are starting these plants from seed. And why not? After eating the fruit, all you have to do is clean the seeds and put them to work. However, none of these exotic fruits is easy to grow; they take some coddling. So attempt them only after you have been successful with the other plants mentioned.

With these plants you must separate the seed from the pulp, and this is not as easy as one might think. But it is not that difficult. You can soak the fruit seeds in water for a day and then remove the old fruit flesh with your fingers. Or do the separation process in a jar of water; the pulp will go to the bottom of the jar while the viable seeds will come to the surface.

Let the seeds dry a day and then sow them in shallow trays—4-inch trays or planters are fine—in vermiculite. Put the trays in a warm place (75 F.) and cover them with some plastic. Keep the vermiculite evenly moist. Germination may occur in a few weeks or a few months or not at all. You have to be patient with tropicals. Once the plants are growing and are 6 to 8 inches high, transplant them into individual pots of houseplant soil.

The extra care and patience to grow these plants are worth it. Not everyone has a papaya or pomegranate in the house.

7
Vegetables and Herbs

Everyone is growing vegetables these days because homegrown produce tastes better than any you buy at the market, is much cheaper, and is sometimes safer (no pesticides on them).

Sowing vegetable seed is so easy that even if you make some mistakes you are bound to get *some* plants and a harvest. The main consideration with vegetable growing is to plant the right vegetables at the right time. Some vegetables, like lettuce, carrots, beets, spinach, and radishes, should be planted when the weather is cool; others, like peppers, beans, melons, eggplant, and tomatoes, need warm weather.

You can buy prestarted plants—seedlings—but it makes more sense to grow your own and costs much less too. And it is the only way to get a head start on spring, since many vegetables can be started indoors.

Warm- and Cool-Season Vegetables

Warm-season crops need heat, and lots of it, for a fairly long time. The temperature should be at least 65 F. if the plants are to thrive. The cool-season crops do well between 50 and 65 F.; these plants grow their best in cool but not cold weather. In addition to warm- and cool-season crops, there are early- and late-producing varieties. The early ones need less heat than the later ones. All these planting considerations are governed by the weather (planting time) and where you live (zone). The time of planting is marked on most seed packages.

The reference guide to growing plants in different areas of the United States is the hardiness zone maps used by nurserymen and gardeners. This is somewhat like a universal guidebook of what you can and cannot grow, and is prepared by the Agricultural Research Service of the U.S. Department of Agriculture. This map is based on the average minimal night temperatures from weather stations, and separate areas of the United States into zones. The map is available by writing to the United States Department of Agriculture in Washington, D.C. (There is a nominal charge.)

Thinning and Transplanting Vegetable Plants

Thinning vegetables is quite important if you are growing the plants from seed. The plants must be thinned so they have growing space between them. If you do not thin them, they cannot grow correctly, and crowded conditions invite bugs and disease. (Food plants started outdoors are more subject to insect damage than when they are started indoors.)

Thin out your growing plants as follows:

Beans	3 to 6 inches apart
Beets	2 to 3 inches apart
Brussels sprouts	10 to 12 inches apart
Cabbage	10 to 12 inches apart
Carrots	2 to 3 inches apart
Cauliflower	10 to 12 inches apart
Eggplant	6 to 8 inches apart
Lettuce	4 to 8 inches apart
Peas	2 inches apart
Peppers	8 to 10 inches apart
Potatoes	6 inches apart
Radishes	1 to 2 inches apart
Spinach	6 inches apart
Squash	15 to 20 inches apart
Tomatoes	10 to 15 inches apart

Let us review the transplanting process. Transplant the seedlings very carefully so you do not injure the plant roots. The less shock small plants get, the better they will grow.

For this reason a one-step planting method deserves consideration, especially with vegetables: sowing seeds directly into biodegradable small blocks or cubes. This means no transplanting (the seeds are directly planted outdoors in the blocks or cubes, and the plants do not have to be removed from these "containers") and less chance of a plant dying from transplant shock. In Chapter 2 we only briefly mentioned the small pellets and cubes. Now let us consider them in more detail:

Jiffy 7 Pellets are compressed pellets of sphagnum peat and soil, with some fertilizer added. When the pellet is watered, it expands. The seed is placed directly into the pellet.

Kys Kubes are a type of fiberboard mixed with fertilizer. The entire cube is planted when the seeds are growing.

BR 8 Blocks are fiber blocks containing fertilizer. When the seedlings reach their desired size, the blocks are planted in permanent places.

Peat pots are perhaps the most popular. These are fiber pots filled with synthetic soil.

Spacing Seeds and Seed Depth
for the Common Vegetables

The following chart includes information on how to space seeds in the outdoor garden, the depth to plant seed, and other pertinent data:

Vegetables: Seed Spacing and Depth, Germination Temperatures and Times

Common Name	Distance Between Plants (in Inches)	Distance Between Rows (in Inches)	Depth to Plant Seed (in Inches)	Germination Temperature (F.)	Germination (in Days)
Artichoke	60	72	½	68–86	7–21
Asparagus	18	36	1½	68–86	7–21
Beans:					
Garden	6–8	18–20	1½–2	68–86	5–8
Lima	6–8	24–30	1½–2	68–86	5–9
Runner	6–8	Grow vertically	1½–2	68–86	5–9
Beet	2	12–14	1	68–86	3–14
Broccoli	12–14	24–30	½	68–86	3–10
Brussels Sprouts	12–18	24–30	½	68–86	3–10
Cabbage	16–20	24–30	½	68–86	3–10
Cabbage, Chinese	12–18	20–24	½	68–86	3–7
Carrot	1–2	12–18	¼	68–86	6–21
Cauliflower	8–10	30–34	½	68–86	3–10
Celery	8–12	24–30	⅛	50–68	10–21
Chard, Swiss	4–8	18–24	1	68–86	3–14
Corn, Sweet	4–6	30–36	2	68–86	4–7
Cress, Garden	10–12	12–16	¼	68	4–10
Cucumber	10	40–50	1	68–86	3–7
Eggplant	12–16	30–36	½	68–86	7–14
Endive	9–12	12–24	½	68–86	5–14
Kale	8–12	18–24	½	68–86	3–10
Kohlrabi	3–4	18–24	½	68–86	3–10
Leek	2–4	12–18	½–1	68	6–14
Lettuce	12–14	18–20	¼–½	68	7

Vegetables: Seed Spacing and Depth, Germination Temperatures and Times

Common Name	Distance Between Plants (in Inches)	Distance Between Rows (in Inches)	Depth to Plant Seed (in Inches)	Germination Temperature (F.)	Germination (in Days)
Muskmelon (including Cantaloupe)	12–16	48–72	1	68–86	4–10
Mustard	2–6	12–18	½–1	68–86	3–7
Okra	15–18	28–30	1	68–86	4–14
Onion	2–3	12–14	½	68	6–10
Parsnip	3–4	16–20	½	68–86	6–28
Pea	2–3	Grow vertically	2	68	5–8
Pepper	16–18	24–30	¼	68–86	6–14
Potato	12–14	24–36	4	68	
Potato, Sweet	12–18	36–48	(Plants)	77	
Pumpkin	30	72–100	1½–2	68–86	4–7
Radish	1–2	6–12	½	68	4–5
Rhubarb	30–36	60	Up to crown	68–86	7–21
Rutabaga	8–12	18–20	½	68–86	3–14
Spinach	2–4	12–14	½	59	7–21
Spinach, New Zealand	16–18	20	2	50–86	5–28
Squash	20–24	Grow vertically	1	68–86	4–7
Tomato	18–30	30–48	½	68–86	5–14
Turnip	1–3	15–18	½	68–86	3–7
Watermelon	14–18	50	1	68–86	4–14

Insects and Disease

You have to keep young plants free of insects and disease so they mature into healthy ones. Young plants are the most tender and inviting to insects, so you should know how to protect your fledglings from attack.

You can do natural preventive planting. For example, plant insect-repelling orange nasturtiums (really herbs) among the crops to thwart aphids; or the fragrant herb tansy, to repel cutworms and cabbageworms.

Besides preventive planting, just pick off the insects. Or put cardboard collars around your young plants to thwart insects; most insects think twice about climbing. You can also use such organic solutions as laundry soap and water, tobacco and water, garlic and water, onions and water, and so on, to deter aphids, mealybugs, and scale. Apply these solutions directly on the bugs. Always wash off your crops almost immediately after applying these solutions.

Another method is to use botanical sprays. These sprays contain natural organic chemicals: pyrethrum is made from chrysanthemum flowers; rotenone is derived from the derris root; and ryania comes from a shrub. Pyrethrum kills aphids, leafhoppers, thrips, and whiteflies. Rotenone will eliminate aphids and pea weevil. Ryania will paralyze cabbage loopers. All these sprays are sold at nurseries.

Botanical sprays should repel most insects, but if the attack is heavy and bugs gain a foothold, you might have to resort to the chemical sprays Diazinon, Malathion, and Sevin. Diazinon is a good preventive against

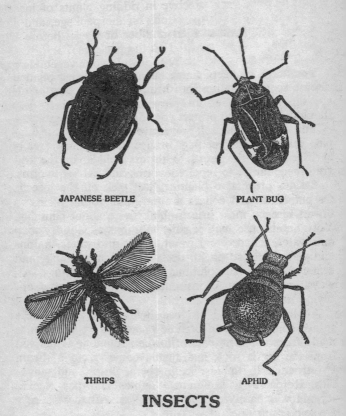

JAPANESE BEETLE   PLANT BUG

THRIPS   APHID

INSECTS

cutworm. Malathion gets rid of aphids, cutworms, and leafhoppers. Sevin takes care of cutworms. These chemicals are poisonous, but they do not accumulate in the soil and are very effective in ridding plants of insects. Always follow the directions on the package, and stop spraying at the specified number of weeks before harvest (this too is indicated on the package).

The insects most likely to invade your vegetable garden are aphids, bean leaf beetles, cutworms, plant bug, leafhoppers, whiteflies, Mexican bean beetles, cabbage maggots, cabbageworms and -loopers, tomato hornworms, pea weevils, pea beetles, and (among non-insects) snails and slugs. There are, also, specific remedies and preparations for tomato hornworms, pea beetles, and snails and slugs at your nursery. For the rest, either handpick the insects, use organic solutions or sprays, or—if absolutely necessary—try the poisonous chemicals.

Diseases too may attack crops, especially young plants. The most common vegetable diseases are mildew, which appears as a white or gray powder on leaves and stems, and rust, which attacks the undersides of leaves with red or brown spores. Blight and scab are other diseases that occasionally invade the vegetable patch.

The best preventative against any disease is to buy disease-resistant varieties of vegetables. But if the plants still suffer from infestation, use chemical fungicides. The two most popular fungicides are Captan and Phaltan; use them according to the directions on the package.

Not a disease or an insect, but sometimes a major problem, is that little varmint the gopher. Smarter than smart and almost impossible to eliminate, the gopher has the same liking for vegetables as people do. Either plant enough vegetables for him, or just curse—I have yet to find an effective gopher repellent.

Herbs

You should consider growing herbs, whether for decoration, medicinal purposes, or food. Little pots of herbs indoors provide a pleasant note of green; and outdoors, as mentioned, some herbs can act as insect repellents. Herbs can also be used as fragrant pomanders or brewed for throat-soothing teas. Finally, everyone is aware of herbs like dill and rosemary, used with foods. Herbs are little in size but big in benefits, so definitely grow some.

You can grow herbs from prestarted plants—these are available in 3-inch pots at nurseries—or start herbs from seed. If you buy the small plants, transplant them to the garden as soon as you get them home. (You can also grow herbs indoors in pots, as explained later.) The tiny amount of soil in their original pots is sparse indeed. Remove the plants from the pots with as much of the root ball intact as possible; dig generous holes and add soil around the plants. Pack the soil down and water the plants thoroughly. Provide a sunny location for herbs; most of them do best in very bright places and few survive shade for long.

If you start your herbs from seed, you can do this indoors in pots of standard starting medium. (Spring or fall is the best time to start seed indoors.) Use 8- or 10-inch-diameter pots and plant three or four seeds to a container; or use plastic trays as described in Chapter 2. Keep the soil evenly moist. Avoid a soggy soil, which can cause the seeds to rot; as well as dry soil, which can starve the seeds. Herbs, like most plants, require good humidity (about 40 to 50 percent to germinate), so

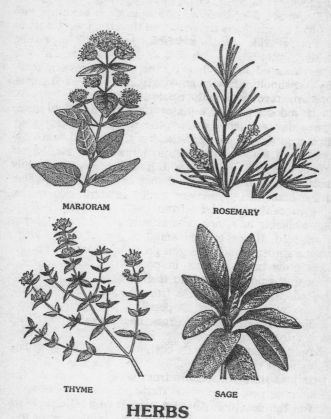

MARJORAM

ROSEMARY

THYME

SAGE

HERBS

cover their containers with plastic bags. Thin out the plants if they become crowded; keep a 2- to 3-inch space between them. Once the plants are 3 to 4 inches tall, remove the plastic and transplant the seedlings into 8- to 10-inch-diameter pots. Use three or four plants to a pot; provide a standard houseplant soil. As a rule, herbs do not need a rich soil, so do not add anything to the packaged soil. When the herbs are growing readily—up another few inches—they can be planted in the garden or grown on in their pots at windows.

If you start herbs outdoors from seeds, plant them when the weather is settled; the best time is spring. A fine seed should be sprinkled on the ground and covered with a thin layer of soil; larger seeds can be set about 1/4 inch under the soil. Keep the seed bed evenly moist. Thin out the plants if they become too crowded. Most herb seed germinates readily in about two weeks.

Herbs to Grow

Following is a list of recommended herbs:

Basil (Ocimum basilicum)

Basil's clovelike flavor is good with cheese, fish, poultry, and tomatoes. Basil needs a fairly rich soil, kept evenly moist, and lots of sun. An annual, it grows to 20 inches.

Borage (Borago officinalis)

The gray-green leaves taste a bit like cucumber. Use borage in salads. Borage needs a slightly poor soil. Let the soil dry out between waterings. An annual, borage has blue flowers.

Chervil (Anthriscus cerefolium)

Chervil has an aniselike flavor. Use it like parsley. Chervil likes a poor, somewhat moist (but not soggy) soil, and sun, although it can tolerate some shade. An annual, it grows to 20 inches.

Chives (Allium schoenoprasum)

Plant chives in fairly rich soil; keep the soil evenly moist. Give these plants lots of sun. Perennials, chives grow to 2 feet. Chives are excellent in salads, soups, and stews.

Dill (Anethum graveolens)

Use dill leaves fresh or dried, or use the seeds in fish and lamb dishes. Dill needs full sun and a well-drained soil. Keep the soil fairly moist. An annual, dill grows to 3½ feet.

Fennel *(Foeniculum vulgare)*

Both fennel leaves and seed have an aniselike flavor. Use fennel in fish and vegetable dishes. Fennel needs full sun and a porous soil. A perennial, it grows to 5 feet.

Lavender *(Lavandula officinalis)*

Use lavender flowers in potpourris. Aromatic lavender must have full sun and a porous, moist, perfectly draining soil. Prune off the flowers after the plant blooms in order to keep the plant compact. A perennial, it grows to 2 feet.

Lovage *(Levisticum officinale)*

Lovage stalks, seeds, and leaves taste and smell like celery. Lovage is delicious in soups, stews, and salads. It needs a moist, slightly alkaline soil and a semi-sunny location. A perennial, it grows to 4½ feet.

Marjoram *(Marjorana hortensis)*

Use fresh or dried leaves of marjoram in casseroles or salads for a tangy treat. Grow marjoram in full sun, and keep the soil evenly moist. A perennial which can be grown as an annual, it reaches 20 inches.

Oregano *(Origanum vulgare)*

Oregano leaves taste a bit like thyme. Fresh or dried leaves are perfect in Spanish and Italian dishes. Oregano needs a well-drained soil, plenty of sun, and the right amount of water (but do not overwater the plants). Cut back the flowers to encourage more flowers. A perennial, it grows to 2 feet.

Parsley *(Petroselinum crispum)*

Use parsley fresh or dried in many dishes. Parsley is an excellent source of protein. Grow it in sun or partial sun and a moist soil. A biennial or perennial, it grows to 12 inches.

Rosemary *(Rosmarinus officinalis)*

Use fresh or dried rosemary leaves with chicken, meat, and vegetable dishes. Grow this herb in poor but well-drained soil in a hot and sunny location. A perennial, it grows to 4 feet.

Sage *(Salvia species)*

Use the fresh or dried gray-green leaves with lamb and meat stuffings. Grow sage in a poor but well-drained soil and full sun. If you overwater the plants, mildew will hit. Cut back the stems after blooming in order to encourage branching. A perennial, it grows to 2½ feet.

Savory (*Satureia species*)

The annual is summer savory (*S. hortensis*); winter savory, a perennial, is *S. montana*. Use the pepper-flavored savory leaves fresh or dried with meats, fish, and eggs and in salad dressings. Savory likes a sandy, well-drained soil, and average moisture, neither too dry nor too wet. Keep the stems clipped. Grows to 18 inches.

Tarragon (*Artemisia dracunculus*)

Tarragon leaves, which have a tangy flavor, can be used dried or fresh in salads and with cheese and fish. Tarragon needs a well-drained, good and rich garden soil, and partial sun. A perennial, it grows to 20 inches.

Thyme (*Thymus vulgaris*)

The gray-green leaves can be used dried or fresh in vegetable, poultry, and meat dishes. Thyme needs a light, warm location and excellent drainage. There are caraway- and lemon-scented thyme varieties. A perennial, it grows to 12 inches.

8
Annuals and Perennials

Annuals and perennials are the flowers of outdoor gardens, but some, like nasturtiums and marigolds, can be grown indoors. You can buy most annuals and perennials as prestarts at nurseries. But if you have any amount of ground to cover, prestarts are expensive. On the other hand, sowing your own seed and growing your own plants is inexpensive, provides you much satisfaction, and lets you grow what you want, not just what is available at local nurseries. A fourth advantage is that generally seeds you germinate are superior to any prestarts: when you grow them yourself, you give them more care than when they are grown for nursery sale, where many times they are grown too rapidly.

Starting annuals and perennials from seed used to be tricky, but today anyone can do it by using the many improved varieties. Seed sowing was covered in Chapter 2, so here we will only review it briefly and discuss how to grow on the annual and perennial seedlings once they are ready for permanent transplanting outdoors.

71

AGERATUM

PORTULACA

NASTURTIUM

PETUNIA

ANNUAL SEEDLINGS

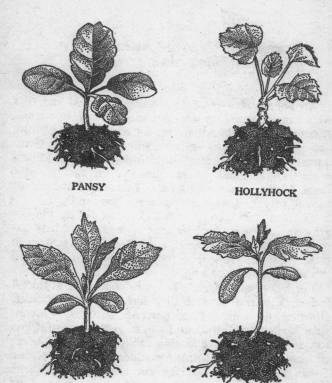

PANSY

HOLLYHOCK

ASTER

PAINTED DAISY

PERENNIAL SEEDLINGS

read on the 33rd frost date in spring for your general area. You can get information by calling your Agricultural Experiment...

How to Start Annuals and Perennials Indoors

An annual is a plant that grows, blooms, and dies within one season. Petunias are examples of an annual. A perennial is a plant that lives for many years; it grows new again every year—without your replanting it. Geraniums are popular perennials. Between annuals and perennials are biennials, plants that live for two or three years. Hollyhocks are biennials. There are hundreds of annual, perennial, and biennial seeds you can buy.

You can start seed indoors in window boxes, window greenhouses, or in the basement under lights. Many annuals and perennials need a long growing season to make complete growth; so they must be sown early and this means indoor growing.

For indoor starting, use boxes or containers at least 4 inches deep and be sure they have drainage facilities. However, most annual and perennial seeds need a constant temperature of about 78 F. to germinate, so if you sow seed indoors, put heating cables in the seed boxes or containers or put the containers in warm places. Some plants—vegetables for example—require precise seed depths to germinate, but with most annuals and perennials, seeds can be sown even one half their normal depth in the starting medium. If the seeds are very fine, then just scatter them on the medium. Space the seeds as you want. You can always thin them out if they are too crowded. Use a sterile growing medium as outlined in Chapter 2.

Planting time for annual and perennial seeds varies over the United States. If you are planting indoors this is not a prime factor, but outdoors you will have to depend on the last frost date in spring for your general area. You can get this information by calling your Agricultural Experiment Station.

Growing On

Bringing annual and perennial seedlings to mature plants is the crucial stage of growing. What you do with the plants after they sprout their first true leaves is the important time, because this is when the youngsters are most vulnerable to mishandling, insects, and disease.

Once you see the first true leaves (usually the second set), transplant the seedlings into individual containers of soil and give them a little bit of plant food. The tiny seedling roots are quite fragile, so they must be handled carefully. Use a blunt-edged wooden stick and *gently* pick the plant out of the starting medium so you get all the roots. You have prepared your new container with fine porous soil; that is, soil which has been sieved of all large lumps. Now firm the soil ever so gently around the plants, and mist them thoroughly or bottom-water them. You must keep the seedlings evenly moist, never dry and never wet.

When the transplanted seedlings are 3 to 4 inches high (this varies with the type of plant), start acclimatizing them to outdoor temperatures as described in Chapter 3.

Once the plants are in the ground, hope and pray there is no frost; frost can wipe out plants. Outside, be sure you give the new plants some extra care: water them thoroughly, remove weeds from around them, and apply some light feeding (10–10–5 plant food) every

DIANTHUS

ZINNIA

NASTURTIUM

CANDYTUFT

ANNUALS

ASTER

ORIENTAL POPPY

PHLOX

CHRYSANTHEMUM

PERENNIALS

third watering. Do not feed the plants too much, because youngsters cannot absorb too much food. Too much food will hurt young plants.

For the most part, both annuals and perennials are sun lovers and to grow them in shade is really defeating the purpose. You may have some bloom but it will be sparse. Try to find an outdoor location that affords at least four hours of sun—though six hours is even better.

Outdoor Sowing

When you start seeds outdoors, directly in the ground, the main consideration is weather; that is, to get the seed into the ground when the weather is settled. As mentioned, this will vary from region to region. In some areas you can start sowing seed in April, in other places not until mid-June, so some discretion must be advised on your part.

It is not necessary to sow the seed of annuals and perennials in rows, as for most vegetables. You can scatter the seed where you want the plants to grow, and if they get too crowded merely thin them out later. However, you will need the same type of good porous soil for growing flowering plants as for vegetables.

As explained in Chapter 3, prepare the beds by digging down at least 15 to 18 inches and enrich the soil with compost. A loose, friable soil is what you need for plants—so seed can germinate readily and so that plants can grow on to healthy mature specimens.

Annuals and Perennials to Grow

The following pages list, by common name first, the perennials and annuals I recommend; where to sow them; and seed-germination times. Plants are listed by (1) common name or (2) genus (or group) name if they have no common name. Genera contain many species.

Annuals and Perennials: Seed Sowing and Germination Times

Common (and Botanical) Name	Type of Plant*	Sow Seed Outdoors After Last Frost	Sow Seed Indoors	Germination (in Days)
Ageratum	A		×	21
Alyssum	P	×		14
Amaranthus	A	×		14
Anemone	P	×		14–21
Arabis	P	×		21–28
Arctotis	A		×	8–10
Argemone	A	×		14–21
Armeria	P	×		21–28
Aster, China (*Callistephus chinensis*)	A		×	14–21
Aubrietia	P	×		14–21
Baby's Breath (*Gypsophila paniculata*)	A & P	×		7–14
Balsam (*Impatiens balsamina*)	A		×	14–21
Bee Balm (*Monarda didyma*)	P		×	7–10
Begonia	P		×	14–21
Bells of Ireland (*Molucella laevis*)	A		×	14–21

* A = annual; P = perennial.

Annuals and Perennials: Seed Sowing and Germination
Times

Common (and Botanical) Name	Type of Plant*	Sow Seed Outdoors After Last Frost	Sow Seed Indoors	Germination (in Days)
Blanket Flower (*Gaillardia aristata*)	P	×		14–21
Blue Lace Flower (*Trachymene caerulea*)	A		×	7–10
Boltonia	P		×	14–21
Browallia (*Browallia viscosa*)	A		×	14–21
Butterfly Weed (*Asclepias tuberosa*)	P	×		10–14
Calendula (*Calendula officinalis*)	A		×	14–21
California Tree Poppy (*Romneya coulteri*)	P		×	14–20
Calliopsis (*Coreopsis tinctoria*)	A		×	14–28
Candytuft (*Iberis sempervirens*)	P	×		14–21

* A = annual; P = perennial.

Annuals and Perennials: Seed Sowing and Germination
Times

Common (and Botanical(Name	Type of Plant*	Sow Seed Out-doors After Last Frost	Sow Seed Indoors	Germina-tion (in Days)
Canterbury Bells (*Campanula medium*)	A		×	5–7
Cerastium (*Cerastium tomentosum*)	P	×		14–28
Chinese Bellflower (*Platycodon grandiflorum*)	P		×	14–21
Chinese Pink (*Dianthus chinensis*)	A		×	7–10
Chrysanthemum (*Chrysanthemum carinatum*)	P		×	14–35
Columbine (*Aquilegia*)	P	×		21–28
Coreopsis (*Coreopsis grandiflora*)	P	×		14–21
Cornflower (*Centaurea*)	P	×		10–15
Cosmos (*Cosmos bipinnatus*)	A & P		×	14–28

* A = annual; P = perennial.

Annuals and Perennials: Seed Sowing and Germination Times

Common (and Botanical) Name	Type of Plant*	Sow Seed Outdoors After Last Frost	Sow Seed Indoors	Germination (in Days)
Dahlia (*Dahlia pinnata*)	A		×	14–21
Datura (*Datura suaveolens*)	A		×	14–21
Delphinium (*Delphinium elatum*)	P	×		14–28
Dimorphotheca	A		×	10–15
Dodecatheon	P		×	7–14
Dusty Miller (*Centaurea gymnocarpa* or *Senecio cineraria*)	A		×	14–21
English Daisy (*Bellis perennis*)	P	×		7–14
Feverfew (*Chrysanthemum parthenium*)	P	×		14–21
Fleabane (*Erigeron*)	P		×	14–21

* A = annual; P = perennial.

Annuals and Perennials: Seed Sowing and Germination Times

Common (and Botanical) Name	Type of Plant*	Sow Seed Outdoors After Last Frost	Sow Seed Indoors	Germination (in Days)
Forget-Me-Not (*Myosotis*)	P	×		21–28
Foxglove (*Digitalis purpurea*)	P	×		14–21
Gaillardia (*Gaillardia picta*)	A		×	14–21
Gas Plant (*Dictamnus*)	P		×	7–14
Gerbera	P	×		20–30
Geum	P	×		20–30
Globe Amaranth (*Gomphrena globosa*)	A	×		14–21
Godetia	A	×		14–21
Helenium	P		×	10–15
Helianthemum	P		×	10–21
Heliopsis	P		×	4–7
Hollyhock (*Altheae rosea*)	A	×		14–21

* A = annual; P = perennial.

Annuals and Perennials: Seed Sowing and Germination Times

Common (and Botanical) Name	Type of Plant*	Sow Seed Outdoors After Last Frost	Sow Seed Indoors	Germination (in Days)
Italian Bellflower (*Campanula isophylla*)	P	×		10–15
Kochia	A		×	15–20
Larkspur (*Delphinium ajacis*)	A	×		21–28
Leopard's Bane (*Doronicum caucasicum*)	P	×		14–21
Liatris	P	×		7–14
Limonium	P	×		7–14
Linaria (*Linaria maroccana*)	A		×	14–21
Lobelia, Blue (*Lobelia erinus*)	A		×	14–21
Lobelia, Red (*Lobelia cardinalis*)	P	×		7–10
Lupine (*Lupinus polyphyllus*)	P & A	×		21–28

* A = annuals; P = perennial.

Annuals and Perennials: Seed Sowing and Germination
Times

Common (and Botanical) Name	Type of Plant*	Sow Seed Outdoors After Last Frost	Sow Seed Indoors	Germination (in Days)
Marigold (Tagetes species)	A		×	7–14
Mignonette (Reseda odorata)	A	×		14–21
Morning Glory (Ipomoea purpurea)	A		×	21–28
Nasturtium (Tropaeolum majus)	A	×		14–21
Nemesia	A		×	7–14
Nicotiana	A & P		×	14–18
Nierembergia	A		×	10–14
Oenothera	P	×		15–20
Pansy (Viola tricolor)	A	×		14–21
Patience Plant (Impatiens)	A		×	14–18
Penstemon	P	×		14–20
Petunia (Petunia hybrida)	A		×	7–14

* A = annuals; P = perennial.

Annuals and Perennials: Seed Sowing and Germination
Times

Common (and Botanical) Name	Type of Plant*	Sow Seed Outdoors After Last Frost	Sow Seed Indoors	Germination (in Days)
Phlox (*Phlox drummondii*)	A		×	14–21
Pinks (*Dianthus deltoides*)	P	×		10–25
Polemonium	P		×	12–15
Poppy (*Papaver orientale*)	P		×	7–14
Primrose (Primula)	P		×	21–30
Pyrethrum (*Chrysanthemum coccineum*)	P	×		12–15
Red-Hot Poker (*Kniphofia aloides*)	P	×		21–28
Rose Mallow (*Hibiscus moscheutos*)	P		×	18–24
Rudbeckia	A		×	20–25
Salpiglossis (*Salpiglossis sinuata*)	A		×	14
Salvia	A		×	15–20

* A = annuals; P = perennial.

Annuals and Perennials: Seed Sowing and Germination
Times

Common (and Botanical) Name	Type of Plant*	Sow Seed Outdoors After Last Frost	Sow Seed Indoors	Germination (in Days)
Scabiosa	A		×	7–10
Schizanthus	A		×	15–20
Shasta Daisy (*Chrysanthemum maximum*)	P	×		15–20
Sidalcea	P		×	12–15
Snapdragon (*Antirrhinum majus*)	A		×	7–14
Speedwell (*Veronica*)	P	×		15–20
Stock (*Mathiola incana*)	A		×	14
Stokesia	P		×	15–20
Strawflower (*Helichrysum bracteatum*)	A		×	7–10
Sunflower (*Helianthus*)	P	×		14–21
Swan River Daisy (*Brachycome iberidifolia*)	A		×	7–10
Sweet Alyssum (*Lobularia maritima*)	A		×	14–21

* A = annual; P = perennial.

Annuals and Perennials: Seed Sowing and Germination
Times

Common (and Botanical) Name	Type of Plant*	Sow Seed Outdoors After Last Frost	Sow Seed Indoors	Germination (in Days)
Sweet Rocket (*Hesperis matronalis*)	P	×		20–30
Thunbergia	A		×	8–10
Tithonia (*Tithonia rotundifolia*)	P		×	14–21
Torenia	A		×	14–21
Trollius	P	×		21–30
Verbena (*Verbena hortensis*)	A		×	21–28
Vinca rosea	P		×	14–21
Violet (*Viola cornuta*)	P	×		14–21
Wallflower (*Cheiranthus cheiri*)	P	×		10–15
Zinnia (*Zinnia elegans*)	A		×	4–7

* A = annuals; P = perennial.

9

Trees and Shrubs

You can grow your own trees and shrubs from seeds *if* you are patient and can tolerate a few failures along the way. Tree and shrub seeds are fleshy or hard-shelled, sometimes winged, or very fine.

The protective seed shell ensures that seeds do germinate, in one manner or another. In nature the weather takes care of most requirements of special seed—for example, hard-coated seeds—but when we cultivate seeds in an artificial environment, we ourselves must duplicate these natural conditions. You will have to give some seeds a period of ripening in moisture at a low temperature. Other seeds will need a period of ripening at high temperatures before being given very cold temperatures. Only by simulating natural conditions as much as possible can you be successful in germinating the seeds of various trees and shrubs. Other than having to *prepare* tree and shrub seed, the actual seed sowing is the same as for the other plants discussed in this book.

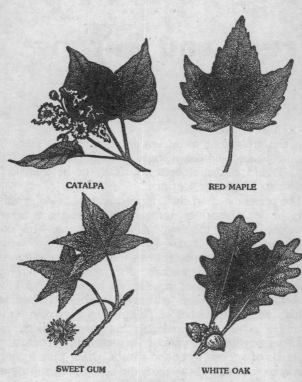

CATALPA

RED MAPLE

SWEET GUM

WHITE OAK

TREES

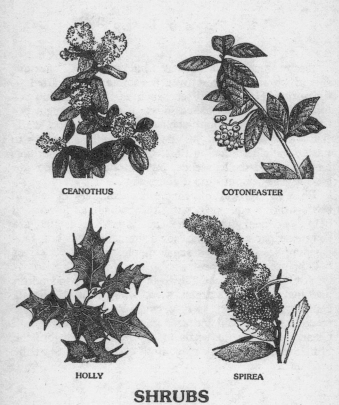

CEANOTHUS

COTONEASTER

HOLLY

SPIREA

SHRUBS

Types of Seeds

Fleshy seed such as oak, buckeye, and chestnut should be stored moist over the winter and started in the spring. Hard-coated seed such as viburnum and cotoneaster should be soaked in warm water overnight before sowing. Some hard-shelled seeds should be scarified—that is, nicked at the edges with a knife—to hasten their germination time. Outside, the weather softens the hard-shelled seeds, but indoors when you are sowing seed you must do it yourself.

Pine, fir, spruce, and maple are examples of winged seeds—those equipped with tiny winglike devices to help float them through the air. Winged seeds that are summer-ripening can be sown immediately, but fall-ripening seeds, such as sugar maple, need a period of cold before they can be sown.

You have to collect fine seed, such as rhododendron and hydrangea and azalea seed, while it is still in its capsule; once it starts to disperse, it is almost impossible to collect. You will find seed capsules in the early fall. Clean and place the seeds in dry bottles and store the bottles inside in a dark place until early spring. Start the seed then.

Stratficiation of Seeds for the Major Trees and Shrubs

Stratification is a ripening process. The tree or shrub seed goes through a period of ripening at low temperatures for 60 to 120 days. Stratification of seed occurs naturally outdoors. The term "stratified" evolved through the years because people used to put the seeds between alternate layers of sand in boxes in a cool place. An easy way now for the home gardener to simulate ripening conditions is to mix the seeds with sphagnum. Put this mixture into plastic bags, put the bags in the refrigerator, and store them there at 41 to 45 F.

The lists of trees and shrubs that follow will tell you (under "Germination Notes") how long and at what temperature to stratify your seeds:

Trees: Flowering and Seed-Dispersal Times, and Germination Notes

Common and Botanical Names	Flowering	Seed Dispersal	Germination Notes*
Alder, Red (*Alnus rubra*)	early spring	November–December	a
Ash, Green (*Fraxinus pennsylvanica*)	May	October–May	a
Ash, White (*F. americana*)	April–May	September–December	a
Basswood (*Tilia americana*)	June–July	fall–spring	b
Beech, American (*Fagus grandifolia*)	April–May	after first heavy frost	a

* Key to "Germination Notes":
a = stratify 2 to 3 months at 32 to 45 F.
b = soak seed in hot water, then stratify for 3 to 5 months at 32 to 45 F.
c = stratify for 2 to 4 months at 68 to 80 F., then stratify for 2 to 4 months at 32 to 45 F.
d = scarify seed coat

Trees: Flowering and Seed-Dispersal Times, and Germination Notes

Common and Botanical Names	Flowering	Seed Dispersal	Germination Notes*
Birch, Gray (*Betula populifolia*)	April–May	October–January	a
Birch, Paper (*B. papyrifera*)	April–June	September–April	a
Birch, Sweet (*B. lenta*)	April–May	September–November	a
Box Elder (*Acer negundo*)	March–May	September–March	a
Douglas Fir (*Pseudotsuga menziesii*)	spring–summer	August–September	a
Fir, Balsam (*Abies balsamea*)	May	September–November	a
Fir, California Red (*A. magnifica*)	June	September–October	a
Fir, Pacific Silver (*A. amabilis*)	spring	October	a

Trees: Flowering and Seed-Dispersal Times, and Germination Notes

Common and Botanical Names	Flowering	Seed Dispersal	Germination Notes*
Fir, White (*A. concolor*)	May–June	September–October	a
Hackberry (*Celtis occidentalis*)	April–May	October–winter	a
Hemlock, Eastern (*Tsuga canadensis*)	May–June	September–winter	a
Hemlock, Western (*T. heterophylla*)	spring	September	a
Hickory, Mockernut (*Carya tomentosa*)	April–May	September–December	c
Hickory, Pignut (*C. glabra*)	April–May	September–December	c
Hickory, Shagbark (*C. ovata*)	April–June	September–December	c
Hickory, Shellbark (*C. laciniosa*)	April–June	September–December	c

Trees: Flowering and Seed-Dispersal Times, and Germination Notes

Common and Botanical Names	Flowering	Seed Dispersal	Germination Notes*
Honey locust (*Gleditsia triacanthos*)	May–June	September–February	d
Larch, Western (*Larix occidentalis*)	spring	August–September	a
Locust, Black (*Robinia pseudo-acacia*)	May–June	September–April	d
Maple, Red (*Acer rubrum*)	February–May	April–July	a
Maple, Silver (*A. saccharinum*)	February–April	April–June	a
Maple, Sugar (*A. saccharum*)	March–May	October–December	a
Oak, Black (*Quercus velutina*)	April–May	September–November	a

Trees: Flowering and Seed-Dispersal Times, and Germination Notes

Common and Botanical Names	Flowering	Seed Dispersal	Germination Notes*
Oak, Northern Red (*Q. rubra*)	April–May	September–October	a
Oak, Scarlet (*Q. coccinea*)	April–May	September–October	a
Oak, Southern Red (*Q. falcata*)	April–May	September–October	a
Pecan (*Carya illinoensis*)	March–May	September–October	a
Pine, Eastern White (*Pinus strobus*)	April–June	September–October	a
Pine, Jack (*P. banksiana*)	May	fall–several years	a
Pine, Loblolly (*P. taeda*)	March–April	fall–spring	a
Pine, Ponderosa (*P. ponderosa*)	April–June	fall–spring	a

Trees: Flowering and Seed-Dispersal Times, and Germination Notes

Common and Botanical Names	Flowering	Seed Dispersal	Germination Notes*
Pine, Western White (*P. monticola*)	spring	fall–spring	a
Red Cedar, Eastern (*Juniperus virginiana*)	March–May	February–March	c
Red Cedar, Western (*Thuja plicata*)	April	August–October	c
Redwood (*Sequoia sempervirens*)	November–March	fall	a
Spruce, Black (*Picea mariana*)	May–June	October	a
Spruce, Red (*P. rubens*)	April–May	September	a
Spruce, White (*P. glauca*)	May	August–November	a

Trees: Flowering and Seed-Dispersal Times, and Germination Notes

Common and Botanical Names	Flowering	Seed Dispersal	Germination Notes*
Sweet Gum (*Liquidambar styraciflua*)	March–May	September–November	a
Sycamore, American (*Platanus occidentalis*)	May	September–May	a
Walnut, Black (*Juglans nigra*)	May–June	fall	a
Yellow Poplar (*Liriodendron tulipifera*)	April–June	October–January	a

Shrubs: Seed-Dispersal Times and Other Data

Shrub	Seed Dispersal	Remarks
Barberry	fall or spring	Stratify 2 to 6 weeks at 40 F.
Camellia	fall	Sow before seed coat hardens.
Ceanothus	spring or fall	Soak seed in hot water.
Cotoneaster	spring or fall	Stratify 3 to 4 months at 70 F., then stratify for 3 months at 32 F.
Euonymus	late fall	Stratify 3 to 4 months at 32 to 50 F.
Holly	spring	Stratify 3 to 4 months at 32 to 50 F.
Oleander	late fall	Plant immediately.
Pittosporum	spring	Soak seed in hot water (2 hours).
Plumbago	late winter	Germinates easily.
Privet	late fall	Stratify for 2 to 3 months at 45 F.
Roses	spring or fall	Stratify 3 to 6 months at 35 F.
Spirea	late summer	Germinates easily.
Viburnum	fall	Stratify 2 to 6 months at 40 F., then stratify 2 to 4 months at 70 to 80 F.

10
Wildflowers

Wildflowers are nature's beauty untouched by people. The delights of jack-in-the-pulpit, Dutchman's-breeches, bloodroot, and other beautiful flowers are well known to people. Yet we all must protect these outdoor beauties so future generations may enjoy them. We cannot take plants from the land (in fact, most states prohibit this), but we can still collect the seeds or buy them—several seed companies now offer wild-flower seed.

Wildflowers are difficult but not impossible to grow. Just follow the advice in this chapter and you will eventually have delightful and unusual wildflowers in your outdoor garden.

Preparing the Wildflower Garden

Most wildflowers like a humusy soil and a protected area where there is some sun but not intense light. A

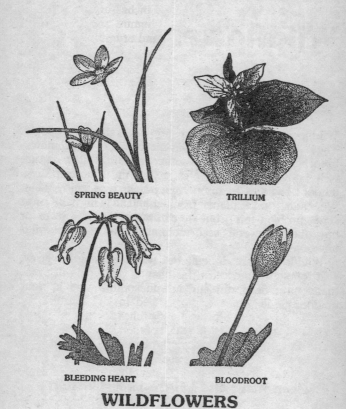

SPRING BEAUTY   TRILLIUM

BLEEDING HEART   BLOODROOT

WILDFLOWERS

place in the shade of deciduous trees is good, because tree leaves protect the plants from heavy summer light, yet allow enough light the rest of the year. Good drainage is also vital; without it wild plants will not grow well. Prepare the soil by removing the top 12 inches and replacing it with rich topsoil. Add some humus to the topsoil, enough to make the soil porous and crumbly. (This varies with the size of the plot.)

Starting Seeds

To break their dormancy, most wildflower seeds need alternate periods of chilling and thawing. In northern regions, start the seeds outdoors in late fall in a rich soil that contains one-third leaf mold and one-third sand. Winter snow will cover and protect the seeds until spring, at which point wet and warm weather will start the plants' germination.

In southern, southeastern, and southwestern regions of the United States, sow the seed in late fall in cold frames, or in containers on porches or in sheltered areas. Again, use equal parts of soil, leaf mold, and sand. Cover the cold frames with a mulch when cold weather starts in order to protect them from rain.

If you used containers (rather than cold frames), barely press *fine* seeds into the medium; cover *large* seeds with 1/4 inch of the mix. Bottom-water the containers—and be sure you do not overwater, or mildew may occur and kill the seeds. Keep the seeds just barely moist through their germination time.

By spring, the seeds should germinate and leaves will begin to develop. When the seedlings are 4 inches high—whether you planted them directly in the soil, in cold frames, or in containers—now transplant them, very carefully, to individual containers; try to take as much of the root ball as possible.

Grow your wildflowers in containers for the first two years of their life, then transfer them to outdoor spots. Do not try to rush the transplanting time—many wildflowers *must* winter over the second year.

Wildflowers to Grow

Wildflowers are lovely native plants and many books have been written about them. There are books for specific states; others are more general. In any case, wildflowers are lovely harbingers of the spring season ahead. If these native plants intrigue you—and they do many gardeners—certainly try your hand at them. Today, many nurseries and suppliers offer the wildflowers of their regions in seed packets. You can also now buy seed from mail-order suppliers. And, of course, you can collect wildflower seed as you do other seed. (See Chapter 1.)

Because there are so many wildflowers, it is difficult to say which are the best; and depending on where you live, the selection you may find or be able to buy will vary. In general terms I have tried to select some of the most commonly grown wild plants that will be adaptable to most soils and climates over a wide area of the United States. However, also consult local books and visit local wildflower shows to see what is growing in your area.

Bloodroot *(Sanguinaria canadensis)*

Bloodroot likes woodsy soil and spring sun, and shade in the summer. The plants bloom in late March, sometimes earlier, with large white flowers.

Blue Phlox *(Phlox divaricata)*

Phlox need a woodsy soil and light shade. Other good varieties are *P. stolonifera* (with lavender flowers), *P. drummondii* (with deep red flowers), *P. longifolia,* and *P. subulata* (ground phlox).

Cardinal Flower *(Lobelia cardinalis)*

Lobelias like moist soil and shade or full sun. Red flowers appear in July to September.

Common American Columbine *(Aquilegia canadensis)*

Most columbines need a dry, tightly packed soil. The plants should have half to full sun. With any luck, your columbine may show its beautiful red flowers the first year, although this is unusual. Also try *A. caerulea* (Colorado columbine), *A. chrysantha* (golden columbine), and *A. flavescens.*

Crested Iris *(Iris cristata)*

Some iris need dryness, but others like rich moist soil and shade. *I. missouriensis* is popular in the Central states; the blue flag, *I. versicolor,* is the favorite in the East. *I. verna* is also desirable.

Dutchman's-Breeches *(Dicentra cucullaria)*

Give these plants shade and a well-drained soil. Dutchman's-breeches grows mainly in Eastern woods, and its foliage dies immediately after flowering. *D. eximia* and *D. formosa* are other good species. The nodding flowers bear two spurs that look like white breeches hanging upside down.

Goldenrod *(Solidago canadensis)*

Goldenrod must have a sandy soil and lots of sun. Do not confuse goldenrod with ragweed!

Hepatica *(Hepatica americana)*

Hepatica grows in a well-drained, somewhat gravely soil. The flowers are pale blue or purple. *H. acutiloba* is another good variety.

Jack-in-the-Pulpit *(Arisaema triphyllum)*

Arisaemas need a rich woodsy soil and shade. Jack-in-the-pulpit usually has striped maroon or pale green flowers, with a white lining. However, other species have different colors. This is an especially easy plant for beginning gardeners.

Large-Flowered Trillium *(Trillium grandiflorum)*

Give these plants shade and keep the soil moist. Trillium has large pink or white flowers.

Shooting Star *(Dodecatheon meadia)*

These plants need a partially shady location and neutral soil. Shooting star is native to the Central and Western states. The Western varieties have red to white flowers.

Spring Beauty *(Claytonia virginica)*

These plants need moist woodsy soil and shade. Lovely pink flowers usually appear in April. *C. virginica* is native to the East, but there are other species native to the West. *C. caroliniana* (broad-leaved spring beauty) is another good plant.

Trout Lily *(Erythronium americanum)*

Trout lilies need deep moist soil and shade, although they can tolerate sun. Native to the Northeast, these easy-to-grow plants have tiny yellow flowers in April. But note: Seed sowing takes almost seven years for blooms!

Violets *(Viola)*

Violets prefer a moist shady spot, but they will tolerate other conditions if they have to. The flowers are yellow to violet to white to red-purple. Yellow species include *V. eriocarpa, V. pubescens,* and *V. rotundifolia.* Blue species include *V. cuculata, V. pedata* (bird's-foot violet), *V. pedatifida,* and *V. sagittata.* White species are *V. blanda* (sweet white violet), *V. canadensis,* and *V. primulifolia.*

Virginia Bluebells *(Mertensia virginica)*

Grow Virginia bluebells in a dry soil, shaded in the summer. After blue flowers appear on 2-foot stems in April, the plants die down completely.

Wild or Wood Geranium *(Geranium maculatum)*

Plant the seedlings in a humusy soil in the shade. These easy-to-grow wildflowers are native to the East and North Central states. You will have pinkish-lavender flowers in April to May.

Wild Ginger *(Asarum canadense)*

Wild ginger needs shade and moist soil. This is a lush, easy-to-grow ground cover.

11
Hybridization

Hybridization is creating outstanding plants that excel in some area—form, fragrance, or color. It is the mating of the best plants with other best plants to produce an even better type of plant. Professional growers spend years and much money to perfect the ideal flower, herb, shrub, tree, or vegetable. But the amateur too can do some plant breeding. This is a delightful and intriguing hobby you might be interested in after you have gained some gardening experience. For commercial growing, plant breeding is an exact science, so a thorough knowledge is necessary in order to produce the ultimate hybrid. But for the average person, extensive genetic data is not necessary. All that is required is some basic information; then you can try your hand at hybridization.

Basic Facts

When you are hybridizing a plant, you have to know something about the plant's parents and grandparents. You must know which characteristics are dominant and which are recessive. These secrets are contained in the genes of the plant.

The genes are living organisms that control the size, growth, color, form, structure, and so on of a plant. There are thousands of male and female genes in a plant, usually in pairs, and these genes determine the appearance of the cross (offspring) when you hybridize two plants. A single gene, or many genes, may control *more than one* or *only one* characteristic at a time. Even when plants are crossed, however, these genes remain independent. Thus, a so-called "recessive"—or un-evidenced—gene may assert itself in a later generation.

Gregor Mendel (1822–1884) is considered the father of genetics because he discovered many laws of heredity, including the law of independent assortment. This law explains what happens when the original parents differ in two or more distinct respects.

By crossing plants, Mendel found that there are fixed laws about certain hereditary characteristics. For example, when dwarf peas were crossed with tall peas, the offspring generation were *not* peas somewhere halfway between tall and dwarf. Instead, the first cross reproduced all tall varieties, with no trace of the dwarf characteristic. Because the dwarf trait did not show up in the first cross, the trait was called "recessive." Mendel called the tall trait "dominant." However, in further studies Mendel found that in the second generation the recessive trait reasserted itself in about 25 percent of the plants. Thus the second generation (called F2) had 25 percent dwarfs and continued to pass on this dwarf trait to future generations. The other 75 percent of the plants were tall, but in future generations they broke up: only one-third of the offspring were tall, and they retained that trait for succeeding years.

This brief description of Mendel's prolific work shows that hybridization is not so much expertise as patience and meticulous record keeping. The chief goal in hybridizing plants is to create several outstanding qualities in one single variety. This can be done only by experimenting with *successive* crossings—you will never do it in one lucky cross.

How to Hybridize

When you transfer pollen from one flower to another flower of the same species, you are hybridizing. The crosses, or transfer of pollen, have to be done on the same species; using plants from totally different families (say, a petunia with a dandelion) will not work. So decide which plant family, or species, you want to work with; and have a definite goal. For example, do you want to create a larger flower, or do you want to get a different color? And do not forget that it takes years to perfect a cross, so be patient and make sure you have the time.

For cross-pollinization, use your thumb or an artist's brush to remove some of the pollen grains from the anthers of a stamen. (Recall from Chapter 1 that the stamens contain the male cells.) Transfer the grains to the stigma of the pistil. Make sure the stigma is sticky and therefore ready to receive the grains. Once the grains are on the stigma, pollen tubes are sent out that extend from the stigma down through the pistil to the ovary chamber below; here the pollen grains mate with the ovules (eggs). The ovules develop into embryos: seeds.

These mechanics are quite simple. It is the timing that is so important. The stigma must be fully formed and sticky, and the pollen must be bursting from the anthers. Sometimes pollen from one of the two plants you intend to cross is ready a few weeks before the pistils of the other plant are ready to receive it. If so, keep the pollen in a tightly capped small dry glass vial. Put some calcium chloride in the bottle, and then add a layer of cotton; place the ripe grains on the cotton. Put the capped vial in the refrigerator.

When you have made the cross, it is extremely important that no other pollen reaches the pistils. Wind or insects could cause self-pollination and ruin the cross, so keep insects away by securely fastening a plastic bag over the flower before it is fully opened.

Helpful Hints

The more you hybridize plants, the more experience you will gain. Stay within one plant family, and be sure of your goal (color, size, or whatever). Select a plant family that interests you and read up on its history and culture. With patience, time, and luck, it is possible to produce an outstanding plant.

For more information about plant hybridization, see the list of books in the Bibliography.

Sources of Seed Supplies

Burgess Seed and Plant Co., Box 1140, Galesburg,
   Michigan 49053

Burpee Seed Co., 370 Burpee Building, Philadelphia,
   Pennsylvania 19132; Clinton, Iowa 52733; Riverside,
   California 92502

Henry Field Seed & Nursery Co., 19 North 12th Street,
   Shenandoah, Iowa 51601

Leslie's Wildflower Nursery, 30 Summer St., Methuen,
   Massachusetts 01844

George W. Park Seed Co., Inc., Greenwood, South Car-
   olina 29547

Pearce Seed Co., Moorestown, New Jersey 08057

Clyde Robin, Box 2091, Castro Valley, California
   94546

Henry Saier, Dimondale, Michigan 48821

R. H. Shumway, Rockford, Illinois 61101

Stoke's Seeds, Inc., 86 Exchange Street, Buffalo, New York 14205

Bibliography

The Complete Book of Growing Plants from Seed, Elda Haring, Hawthorn Books, Inc., New York, New York, 1967

The Genetics of Garden Plants, M. B. Crane and W. J. C. Lawrence, Macmillan & Company, London, England, 1947

Grow Your Own Plants, Jack Kramer, Charles Scribner's, New York, New York, 1973

Handbook on Propagation, Volume 13, No. 2, Sixth Printing, Brooklyn Botanic Garden, Brooklyn, New York, August 1970

Plant Propagation, John P. Mahlstede and Ernest Haber, John Wiley & Sons, Inc., New York, New York, 1957

Plant Propagation in Pictures, Montague Free, Doubleday & Company, Garden City, New York, 1957

Plant Propagation: Principles and Practices, Hudson T. Hartman and Dale E. Kester, Prentice-Hall, Englewood Cliffs, New Jersey, 1959

Plants Under Lights, Jack Kramer, Simon & Schuster, New York, New York, 1974

Practical Plant Propagation, Alfred C. Hottes, A. T. De LaMare Co., Inc., New York, New York, 1925